Groundwood Books / House of Anansi Press
groundwoodbooks.com

We gratefully acknowledge for their financial support of our publishing
program the Canada Council for the Arts, the Ontario Arts Council and
the Government of Canada.

Canada Council   Conseil des Arts
for the Arts   du Canada

ONTARIO ARTS COUNCIL
CONSEIL DES ARTS DE L'ONTARIO
an Ontario government agency
un organisme du gouvernement de l'Ontario

With the participation of the Government of Canada   Canadä
Avec la participation du gouvernement du Canada

Library and Archives Canada Cataloguing in Publication
Title: West Coast wild babies / Deborah Hodge ; pictures by Karen Reczuch.
Names: Hodge, Deborah, author. | Reczuch, Karen, illustrator.
Description: Series statement: West Coast wild
Identifiers: Canadiana (print) 20190146354 | Canadiana (ebook) 20190146362 |
ISBN 9781773062488 (hardcover) | ISBN 9781773062495 (EPUB) |
ISBN 9781773063614 (Kindle)
Subjects: LCSH: Animals—Infancy—Pacific Coast (North America)—Juvenile
literature. | LCSH: Animals—Infancy—Pacific Area—Juvenile literature. | LCSH:
Familial behavior in animals—Pacific Coast (North America)—Juvenile literature.
| LCSH: Familial behavior in animals—Pacific Area—Juvenile literature. | LCSH:
Ecology—Pacific Coast (North America)—Juvenile literature. | LCSH: Ecology—
Pacific Area—Juvenile literature.
Classification: LCC QL151 .H63 2020 | DDC j591.9795—dc23

The illustrations were done in watercolor and color pencil.
Design by Michael Solomon
Printed and bound in Malaysia

MIX
Paper from
responsible sources
FSC
www.fsc.org   FSC® C012700

WEST
COAST
WILD
BABIES

WEST COAST WILD BABIES

DEBORAH HODGE

PICTURES BY

KAREN RECZUCH

GROUNDWOOD BOOKS
HOUSE OF ANANSI PRESS
TORONTO BERKELEY

It's spring on the Pacific west coast, and babies are being born!

All through the ancient rainforest and along the shores of the majestic Pacific Ocean, new life is stirring — high in the trees, low on the forest floor, inside burrows and dens, and deep in the waters of this wild and beautiful place.

Who is being born? Peek into this marvelous wilderness and see.

Gray Wolf Pups

Frisky wolf pups wrestle, chase and pounce,
building muscles and practicing their hunting
skills. When the pups are ready to travel, they will
go to the beach with their pack and learn to dig for
clams, forage for mussels, and fish for salmon.

Black-Tailed Deer Fawn

A newborn fawn is tucked away, quiet and still. The little deer will rest here for a few weeks, hidden among the light and shadows of the rainforest. Being born with almost no scent and a spotted coat helps it stay safe from hungry predators.

Bald Eaglets

Downy gray eaglets hatch high in a nest overlooking the ocean. The parents shred small pieces of fish for the babies to eat. Once their long flying feathers grow in, the young eagles will learn to soar and dive and hunt for their own food.

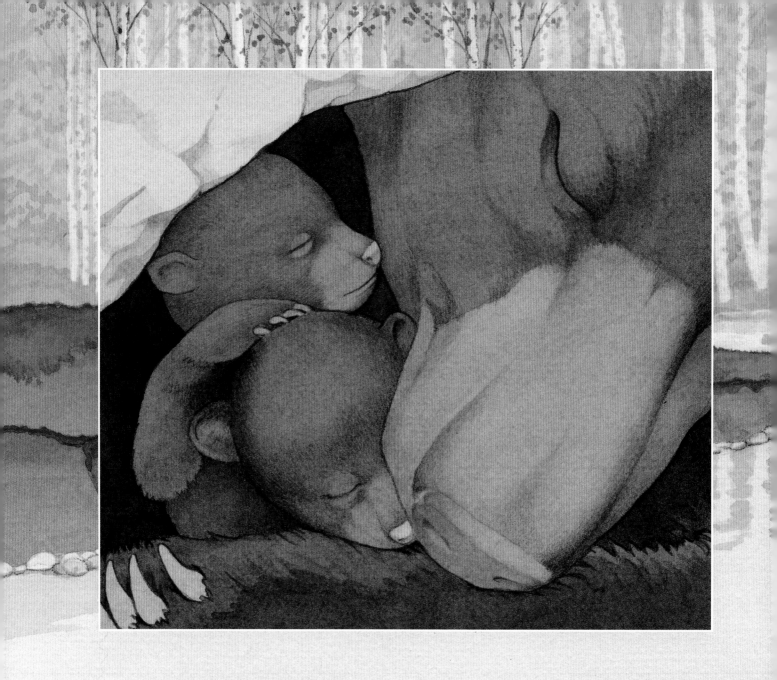

Black Bear Cubs

Tiny cubs are born in a winter den as their mother sleeps. They nestle close and drink her warm milk. By spring, the cubs are the size of furry puppies. Playful and energetic, they frolic and explore the wonders of the forest and seashore.

Orca Calf

Moments after a calf is born, the mother
helps it to the ocean's surface for its first
breath. Soon the baby is breaching and diving
with ease. The lively young whale plays and
swims with its mother and other whales in a
close family group called a pod.

Cougar Kittens

Spotted cougar kittens are born in a den
hidden deep in the rainforest. Purring like
pet cats, they drink their mother's milk.
Soon, they are soft, fluffy bundles of energy.
They stalk, pounce, tumble and play, learning
important skills for hunting.

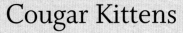

Rufous Hummingbird Chicks

After flying thousands of miles north from her winter home, a mother lays her eggs in a nest made of soft plants and spider silk. Tiny hungry chicks hatch a few weeks later. With her long bill, the mother delivers nectar and insects into their mouths.

Harbor Seal Pup

At the water's edge, a mother gently nuzzles
and nurses her newborn pup. She learns
her baby's scent and sounds. "Maaa," it calls
when its mother is away fishing. The pup
can swim at birth, but it rests on its mother's
back when tired.

Chum Salmon Fry

Every spring, small silver fry emerge from the gravel in their streambeds, where they hatched a month earlier. They swim to the sea, staying near shore at first. Later, they move to the vast open ocean and grow into adult salmon before returning to their home stream to spawn.

Pelagic Cormorant Chicks

Sooty gray chicks hatch on a steep cliff in a nest
made of seaweed and grasses that their parents
have glued to the rock with their guano, or body
waste. With squeaky calls, the babies announce
they are hungry. The parents take turns diving
into the surf to catch fish for the growing family.

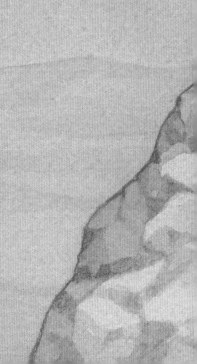

Northern Pacific Treefrog Young

A mother frog lays her eggs in a wetland. After three weeks, tadpoles with long tails hatch and start to swim. Their arms and legs grow, then their tails shrink. Soon the little froglets leave the water to make their home among the forest plants.

Black Oystercatcher Chicks

The parents dig out a nest, called a scrape, on a rocky shoreline and take turns sitting on their speckled eggs. When the fuzzy chicks hatch, they feed on mussels, limpets and other shellfish that their parents capture with their long red beaks.

Sea Otter Pup

A fluffy newborn rides on its mother until it learns to swim and dive. The mother grooms the pup's fur to keep it full of air, helping the baby to float. Before she goes hunting for food, the otter wraps her pup in kelp so that it doesn't drift away.

Gray Whale Calf

In the spring, a calf and its mother travel a great distance along the Pacific coast from Mexico, where the baby was born, to lush feeding grounds in the Arctic. The gentle pair swim close together, making the return journey at summer's end.

Along the Pacific west coast, some babies are born in the ocean, others on shore and still others begin their life deep inside the ancient rainforest.

Whatever their species, and wherever they are born, these baby animals share a magnificent coastal home — a rare and special wilderness to preserve and protect.

The Pacific west coast is a spectacular stretch of land and sea that runs the length of North America, from Mexico to Alaska. Bordering the vast Pacific Ocean, it has some of the most awe-inspiring scenery in the world.

On the west coast of Vancouver Island, in British Columbia, Canada, is the breathtaking Pacific Rim region, which includes Clayoquot Sound and the Pacific Rim National Park Reserve. Situated here are stands of giant ancient trees that comprise some of

the last remaining temperate rainforests on earth. The trees tower above a long sandy beach that meets the immense open ocean. In this magnificent area, a unique and diverse group of wild creatures make their homes and raise their young. All of the species in this book are found in this remarkable region.

Farther north, in the Haida Gwaii archipelago, and along the BC coast to the Alaska panhandle (in an area known as the Great Bear Rainforest), is a similarly vibrant ecosystem. To the south of the

Pacific Rim region is the Pacific Northwest of the United States, with coastal areas of Washington, Oregon and Northern California that also share a forest and marine ecology.

Harbor seals, eagles, whales and salmon are common in many places along the west coast. Other animals, such as bears, cougars and wolves are more frequently found in the northern areas, from BC to Alaska. Some creatures, such as the rufous hummingbird and the gray whale, migrate long distances up and down the coast every year.

The wild babies born on this coastline share a pristine and beautiful home — rare tracts of wilderness that conservationists are working hard to preserve. Keeping the forests intact and the waters clean allows this extraordinary community of wild animals to flourish and grow in a deeply connected web of life.

In the words of the Nuu-chah-nulth First Nations, who have lived as stewards of the land and sea on the west coast of Vancouver Island for thousands of years, "Everything is one and all is connected."

For Further Exploration

Websites
Pacific Rim National Park Reserve
  pc.gc.ca/en/pn-np/bc/pacificrim
(To read about the rainforest, seashore and wildlife, click on "Nature and science")

Pacific Wild
  pacificwild.org
(To see photos and videos of wildlife, the ocean and rainforest, click on "Multimedia")

Books
A Bear's Life by Ian McAllister and Nicholas Read, Orca Book Publishers, Victoria, 2017

Curious Kids Nature Guide: Explore the Amazing Outdoors of the Pacific Northwest by Fiona Cohen, illustrated by Marni Fylling, Little Bigfoot — Sasquatch Books, Seattle, 2017

S is for Salmon: A Pacific Northwest Alphabet by Hannah Viano, Little Bigfoot — Sasquatch Books, Seattle, 2014

Salmon Creek by Annette LeBox, illustrated by Karen Reczuch, Groundwood Books, Toronto, 2004

West Coast Wild: A Nature Alphabet by Deborah Hodge, illustrated by Karen Reczuch, Groundwood Books, Toronto, 2015

Wolf Island by Ian McAllister and Nicholas Read, Orca Book Publishers, Victoria, 2017

Acknowledgments

I would like to express my sincere gratitude to Adrienne Mason, biologist, writer and managing editor of *Hakai Magazine: Coastal Science and Societies*, Tofino, BC. Her thorough review of the manuscript and art was invaluable, as were her answers to many questions about west coast creatures. Thank you also to Bob Hansen, WildSafeBC west coast coordinator, and former human-wildlife co-existence specialist, Pacific Rim National Park Reserve, for his helpful remarks. In addition, I am grateful to my publisher, Semareh Al-Hillal, for her ongoing support, and to the amazing team at Groundwood Books. Special thanks to my wonderful editor, Emma Sakamoto, for her wise and generous counsel; to my co-creator, Karen Reczuch, for her beautiful illustrations; and to Michael Solomon, art director, and Sara Loos, design assistant, for lovely-looking pages. Thank you all!